I COLOR ITALY

ROMA: 753 BC

COLOSSEO: 70 AD

THE COLOSSEUM, AN ANCIENT ROMAN AMPHITHEATER, IS A SYMBOL OF ROME'S IMPERIAL POWER AND ARCHITECTURAL PROWESS. CONSTRUCTION BEGAN UNDER EMPEROR VESPASIAN IN AD 72 AND WAS COMPLETED BY HIS SUCCESSOR, TITUS, IN AD 80. IT COULD SEAT OVER 50,000 SPECTATORS AND HOSTED GLADIATORIAL CONTESTS, PUBLIC SPECTACLES, AND DRAMATIC FIGHTS WITH WILD BEASTS. TODAY, THE COLOSSEUM STANDS AS A TESTAMENT TO ANCIENT ENGINEERING AND IS ONE OF THE MOST VISITED LANDMARKS IN THE WORLD.

FONTANA DI TREVI: 1762 AD

THE FONTANA DI TREVI IN ROME,
A MAJESTIC BAROQUE MASTERPIECE KNOWN WORLDWIDE,
THROW A COIN OVER YOUR SHOULDER FOR LUCK,
WHERE WISHES ARE MADE BENEATH ITS SPARKLING WATERS.

PIAZZA NAVONA

PIAZZA NAVONA IN ROME TRACES ITS ORIGINS BACK TO THE 1ST CENTURY AD, BUILT ON THE SITE OF THE STADIUM OF DOMITIAN, HOSTING ANCIENT ROMAN GAMES,
TRANSFORMED INTO A PUBLIC SQUARE IN THE 15TH CENTURY, EVOLVING OVER TIME,
TODAY, IT STANDS AS A SYMBOL OF BAROQUE ART AND ROMAN HERITAGE.

FOUNDED BY ENZO FERRARI IN 1939, THE FERRARI CAR COMPANY BEGAN AS AN OFFSHOOT OF ALFA ROMEO'S RACING DIVISION. FERRARI GAINED PROMINENCE IN MOTORSPORT, WINNING NUMEROUS CHAMPIONSHIPS. THE ICONIC PRANCING HORSE EMBLEM DEBUTED IN 1947, SYMBOLIZING SPEED AND POWER. TODAY, FERRARI IS SYNONYMOUS WITH LUXURY, PERFORMANCE, AND A RICH RACING HERITAGE.

ITALIAN MEN'S FASHION IS RENOWNED FOR ITS ELEGANCE AND SOPHISTICATION. FROM TAILORED SUITS TO FINE LEATHER SHOES, ITALIANS PRIORITIZE QUALITY AND STYLE. BRANDS LIKE GUCCI, PRADA, AND ARMANI SET TRENDS GLOBALLY WITH THEIR LUXURIOUS DESIGNS. THE ITALIAN MAN EXUDES CONFIDENCE AND FLAIR, EMBODYING A TIMELESS SENSE OF FASHION.

ITALIAN WOMEN'S FASHION IS CELEBRATED FOR ITS ELEGANCE, CRAFTSMANSHIP, AND INNOVATION, SETTING GLOBAL TRENDS WITH ITS TIMELESS APPEAL. AT THE FOREFRONT OF THIS WORLD IS THE RENOWNED STYLIST VALENTINO GARAVANI, SIMPLY KNOWN AS VALENTINO. BORN IN 1932, VALENTINO ROSE TO FAME IN THE 1960S WITH HIS LUXURIOUS DESIGNS THAT EPITOMIZED ITALIAN GLAMOUR. HIS SIGNATURE RED, KNOWN AS "VALENTINO RED," BECAME ICONIC, SYMBOLIZING PASSION AND SOPHISTICATION. VALENTINO'S CREATIONS HAVE DRESSED ROYALTY, CELEBRITIES, AND SOCIALITES, EARNING HIM THE TITLE OF "COUTURIER OF THE CENTURY" BY THE FRENCH GOVERNMENT. HIS FASHION HOUSE, VALENTINO S.P.A., CONTINUES TO BE A SYMBOL OF HIGH FASHION, WITH READY-TO-WEAR AND HAUTE COUTURE COLLECTIONS THAT EMBODY THE ESSENCE OF ITALIAN STYLE. VALENTINO'S INFLUENCE EXTENDS BEYOND FASHION, CONTRIBUTING TO THE CULTURAL AND ARTISTIC LANDSCAPE WITH HIS PHILANTHROPIC EFFORTS AND SUPPORT FOR THE ARTS.

APULIA: THE REGION WHERE IT FEELS LIKE GREECE

SAN PIETRO

THE CHURCH OF ST. PETER IN ROME, A MASTERPIECE OF RENAISSANCE ARCHITECTURE,
DESIGNED BY MICHELANGELO, BERNINI, AND OTHER RENOWNED ARTISTS,
HOME TO MICHELANGELO'S STUNNING PIETÀ SCULPTURE AND THE FAMOUS ST. PETER'S BASILICA,
A SACRED SITE FOR MILLIONS OF PILGRIMS, RICH IN HISTORY AND RELIGIOUS SIGNIFICANCE.

THE FIRST FERRARI CAR, THE 125 S, WAS INTRODUCED IN 1947, MARKING THE BEGINNING OF AN AUTOMOTIVE LEGEND. DESIGNED BY ENZO FERRARI, IT WAS A RACING CAR WITH A V12 ENGINE, SETTING THE TONE FOR FERRARI'S COMMITMENT TO PERFORMANCE AND EXCELLENCE. THE 125 S LAID THE FOUNDATION FOR FERRARI'S SUCCESS IN MOTORSPORTS AND ITS REPUTATION FOR PRODUCING SOME OF THE WORLD'S MOST SOUGHT-AFTER SPORTS CARS. FERRARI'S EARLY MODELS, INCLUDING THE 125 S, ARE NOW HIGHLY PRIZED COLLECTOR'S ITEMS, SYMBOLIZING THE BRAND'S HERITAGE AND PASSION FOR INNOVATION.

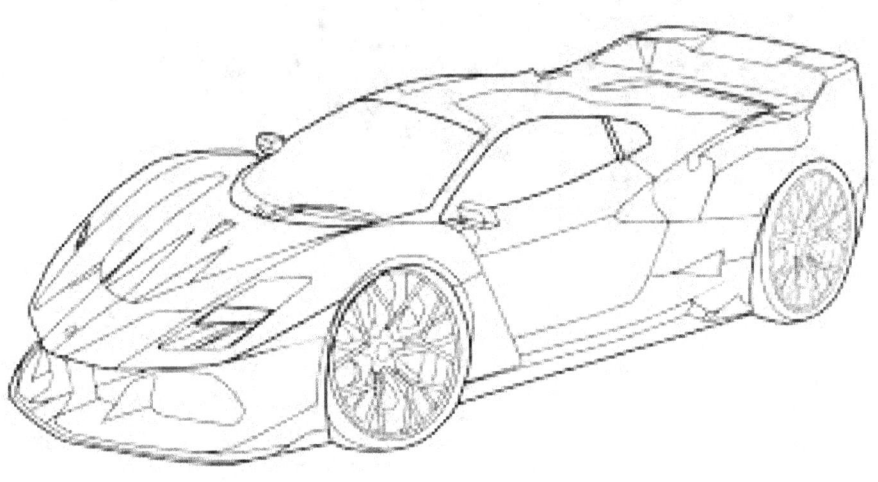

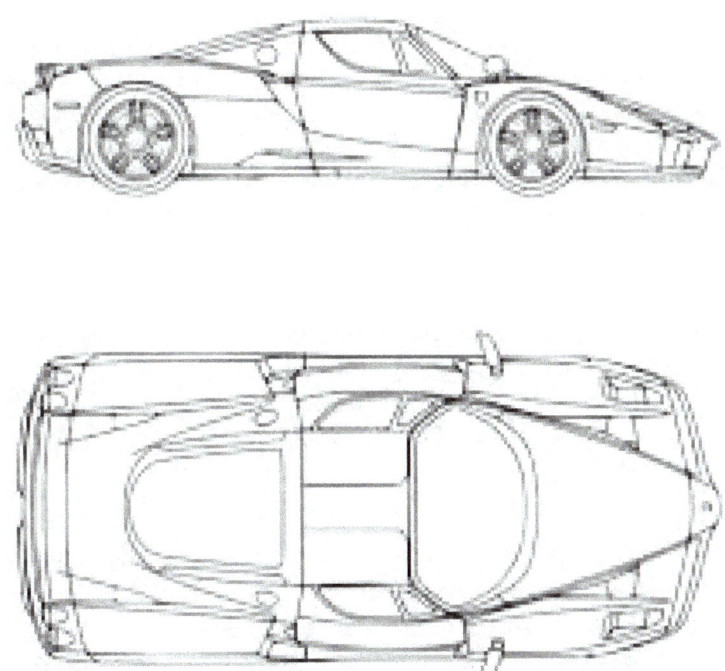

VENEZIA

MONTEPULCIANO

MONTEPULCIANO IS A PICTURESQUE HILL TOWN IN TUSCANY, ITALY, RENOWNED FOR ITS HISTORIC ARCHITECTURE AND WINE PRODUCTION. THE TOWN'S ORIGINS DATE BACK TO THE ETRUSCAN CIVILIZATION, WITH ITS CURRENT LAYOUT LARGELY DEVELOPED DURING THE RENAISSANCE. MONTEPULCIANO IS FAMOUS FOR ITS VINO NOBILE DI MONTEPULCIANO, A PRESTIGIOUS RED WINE THAT HAS BEEN PRODUCED IN THE REGION SINCE THE 14TH CENTURY. THE TOWN'S MAIN STREET, CORSO, IS LINED WITH PALACES, CHURCHES, AND WINE CELLARS, REFLECTING ITS RICH CULTURAL AND ECONOMIC HISTORY. MONTEPULCIANO'S BEAUTY AND CHARM HAVE MADE IT A POPULAR DESTINATION FOR TOURISTS SEEKING AN AUTHENTIC TUSCAN EXPERIENCE.

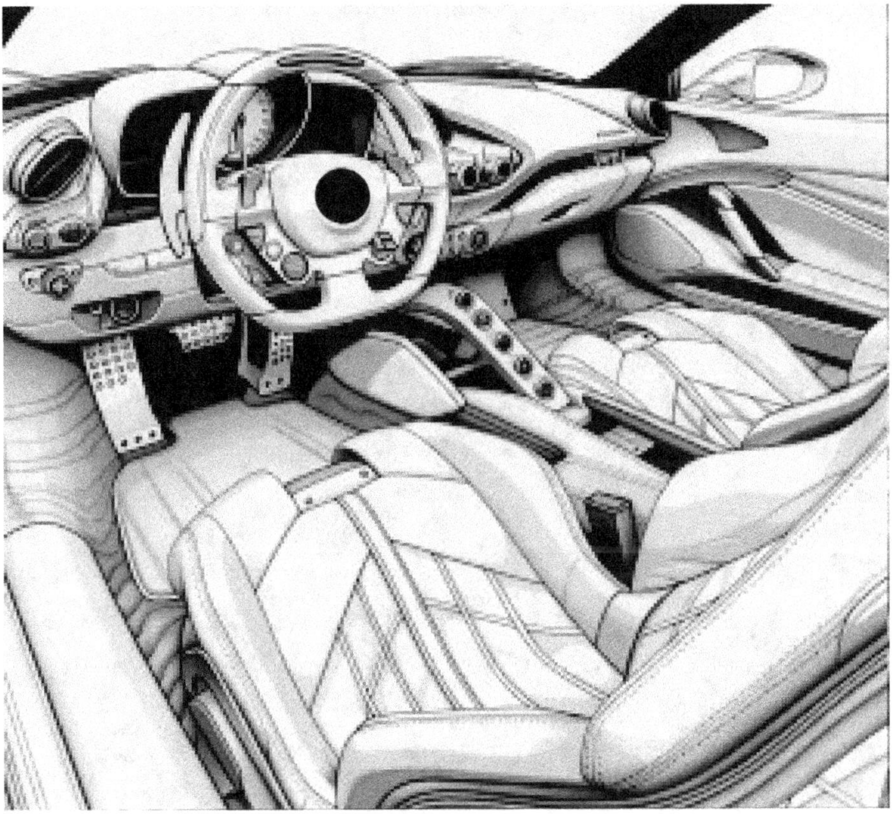

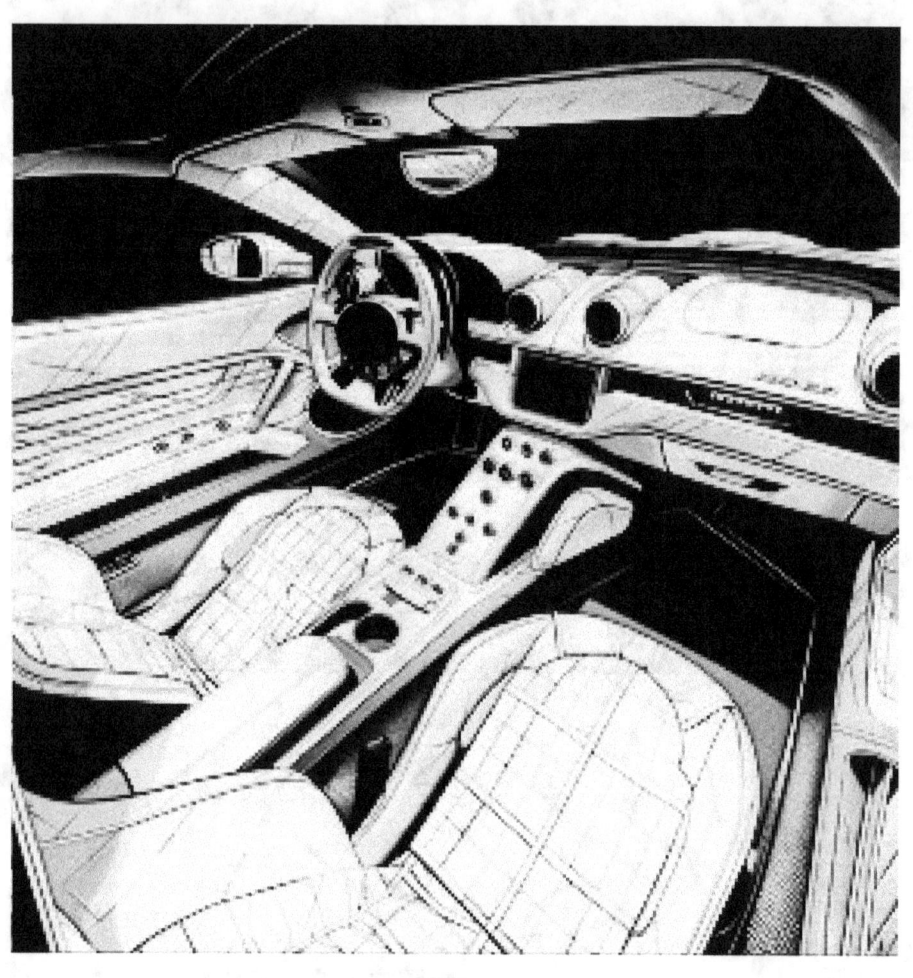

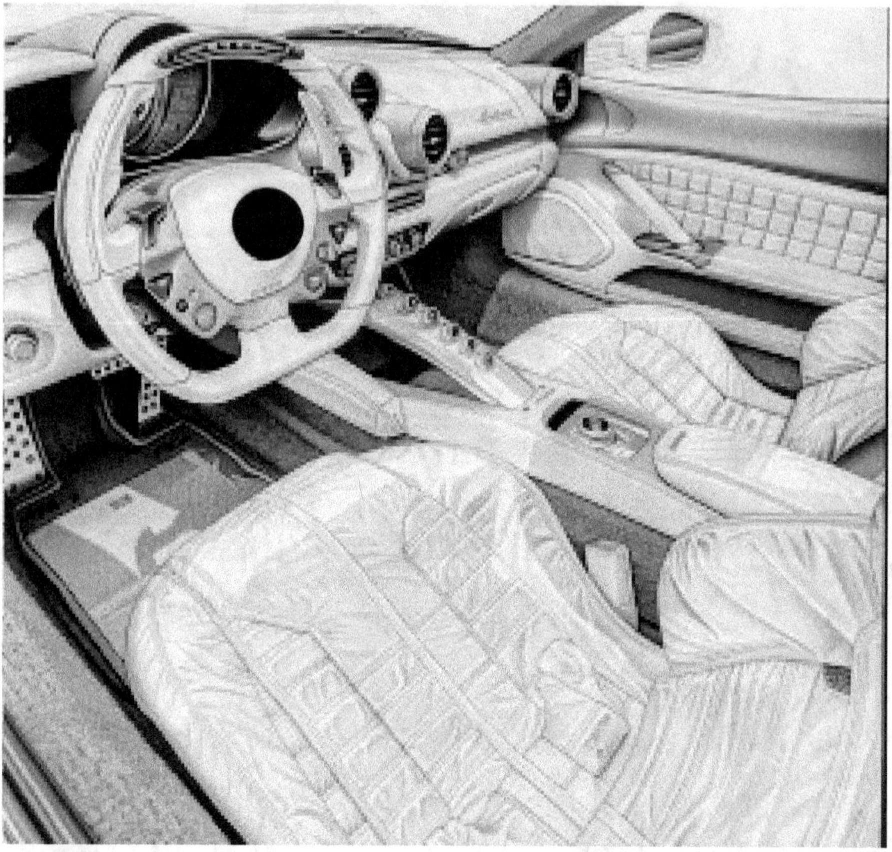

PISA

THE LEANING TOWER OF PISA, AN ICONIC SYMBOL OF ITALY, STANDS IN THE PIAZZA DEI MIRACOLI. BEGUN IN 1173, ITS TILT BEGAN DURING CONSTRUCTION DUE TO SOFT GROUND. WITH A HEIGHT OF 56 METERS, IT LEANS ABOUT 3.9 DEGREES FROM THE VERTICAL. THE TOWER IS ONE OF THE WORLD'S MOST FAMOUS ARCHITECTURAL LANDMARKS, ATTRACTING MILLIONS OF VISITORS.

THE PIAZZA DEI MIRACOLI, OR FIELD OF MIRACLES, IN PISA IS A UNESCO WORLD HERITAGE SITE. IT IS HOME TO SEVERAL ARCHITECTURAL MASTERPIECES, INCLUDING THE LEANING TOWER, THE BAPTISTERY, AND THE PISA CATHEDRAL. THE AREA'S NAME REFLECTS THE AWE AND WONDER INSPIRED BY THESE STRUCTURES. THE PIAZZA IS A TESTAMENT TO THE PISAN ROMANESQUE STYLE, CHARACTERIZED BY ITS USE OF MARBLE AND DISTINCTIVE DESIGN ELEMENTS. IT IS ONE OF ITALY'S MOST VISITED LANDMARKS, ATTRACTING MILLIONS OF TOURISTS WHO COME TO MARVEL AT ITS BEAUTY AND HISTORICAL SIGNIFICANCE.

ROMA: IL CAMPIDOGLIO

THE CAPITOLINE HILL, ONE OF ROME'S SEVEN HILLS, IS A HISTORIC AND CULTURAL HUB. IT IS HOME TO THE CAPITOLINE MUSEUMS, HOUSING ANCIENT ROMAN ART AND ARTIFACTS. THE PIAZZA DEL CAMPIDOGLIO, DESIGNED BY MICHELANGELO, IS A MASTERPIECE OF RENAISSANCE URBAN PLANNING. THE CAPITOLINE HILL OFFERS PANORAMIC VIEWS OF THE ROMAN FORUM AND THE CITY, EMBODYING ROME'S RICH HISTORY AND ARTISTIC HERITAGE.

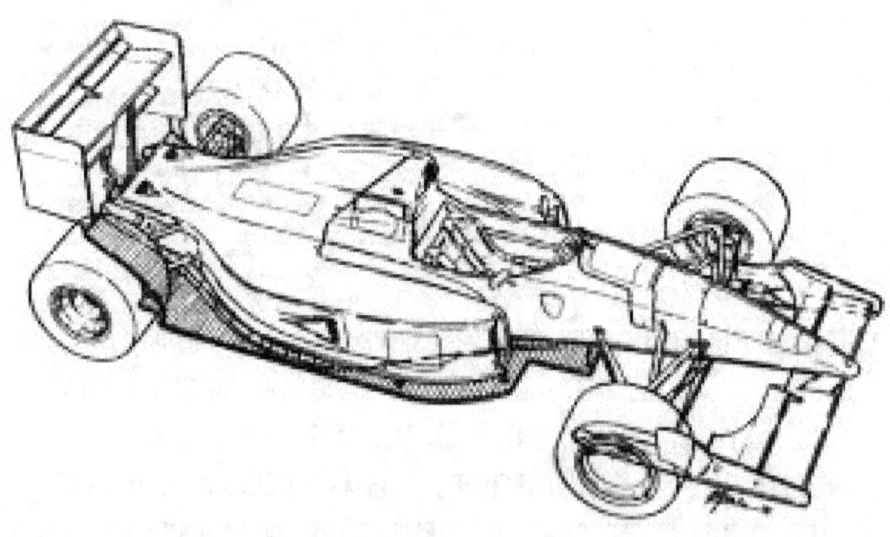

FIRENZE

Florence, the cradle of the Renaissance, is a city steeped in art, history, and culture. Founded by Julius Caesar in 59 BC, it became a Roman colony known as Florentia. During the Middle Ages, Florence flourished as a banking and trade center, laying the groundwork for its cultural renaissance. The city is home to some of the world's most renowned artworks, including Michelangelo's David and Botticelli's Birth of Venus. The Duomo, with its distinctive red dome, is a masterpiece of Renaissance architecture. The Ponte Vecchio, a medieval stone bridge, spans the Arno River, lined with shops that have been there since the 16th century. Florence's influence on art, politics, and science during the Renaissance has left an indelible mark on Western civilization. Its museums, galleries, and historic sites, such as the Uffizi Gallery and the Palazzo Vecchio, continue to attract visitors from around the globe. The city's culinary traditions, including Tuscan wine and cuisine, add to its allure. Florence remains a vibrant cultural hub, blending its rich heritage with modern Italian life.

THE WONDERFUL VIEW OF FLORENCE FROM PIAZZALE MICHELANGELO

FILIPPO BRUNELLESCHI'S DOME, A MASTERPIECE OF RENAISSANCE ARCHITECTURE, CROWNS THE FLORENCE CATHEDRAL. COMPLETED IN 1436, IT WAS A GROUNDBREAKING ACHIEVEMENT IN ENGINEERING AND DESIGN. THE DOME'S INNOVATIVE CONSTRUCTION TECHNIQUE DID NOT REQUIRE A WOODEN SUPPORT STRUCTURE, A FIRST OF ITS KIND. STANDING AT 114.5 METERS, IT REMAINS THE LARGEST MASONRY DOME EVER BUILT. BRUNELLESCHI'S CUPOLA IS A SYMBOL OF FLORENCE'S CULTURAL AND ARTISTIC RENAISSANCE.

Arrivederci!

Goodbye!

www.ingramcontent.com/pod-product-compliance
Lightning Source LLC
Chambersburg PA
CBHW052337220526
45472CB00001B/457